STARS

STARS

JOANNE RIPPIN

PHOTOGRAPHS BY MICHELLE GARRETT

LORENZ BOOKS

LONDON • NEW YORK • SYDNEY • BATH

First published in 1996 by Lorenz Books

Lorenz Books is an imprint of
Anness Publishing Limited
Boundary Row Studios
1 Boundary Row
London SE1 8HP

This edition is distributed in Canada by
Raincoast Books Distribution Limited

ISBN 1 85967 277 9

Publisher: Joanna Lorenz
Introduction by: Beverley Jollands
Designer: Lilian Lindblom
Photographer: Michelle Garrett
Step photography: Janine Hosegood
Illustrations: Lucinda Ganderton

Printed in Singapore by
Star Standard Industries Pte Ltd

CONTENTS

INTRODUCTION

Stars are everywhere. Over the last few years they have covered every surface: walls, clothes, fabrics, armchairs, porcelain, stained glass and jewellery. They have even arrived on the dinner table – as aromatic star anise and slices of carambola, otherwise known as star fruit. And yet their appeal is as strong as ever. Why do we love them so much? Perhaps because a star is such a strong shape: you can't help seeing it as a focal point. Perhaps because their image is so positive, spelling fame and success, exceptional quality and a sense of achievement – we all like to feel we've deserved a star. Real stars are a mystery: infinite in number, infinitely far away, unknown, unattainable. They're also very beautiful. Starlight is delicate and rarefied, blotted out by brighter, man-made lights, but in perfect weather – on a warm summer night or in crisp frosty midwinter – the stars seem to grow larger and ever more numerous.

An Egyptian papyrus of 1000 BC shows the sky-goddess Nut bending over the world, emblazoned from head to foot with regular five-pointed stars. Other ancient ideas of the universe pictured the sky as a vast dome decorated, or perhaps pierced, with stars.

Hundreds of stars glittering in the night sky are echoed in the beautiful star-painted ceilings of some European medieval churches. The fourteenth-century English Court of Star Chamber got its name, so it is said, because the council chamber of the old Palace of Westminster had a ceiling covered with gilded stars. Flat, stylized decoration like this was revived with Victorian Gothic, and gilded stars appeared in interiors by Pugin and Gilbert Scott. It's still a lovely way to use them: small and delicate against a rich dark background, warmly glinting specks of gold. A very successful wallpaper design of recent years has used a simple gold star printed over subtle, rich tones – a pattern much copied with a stencil over colour-washed walls.

The star we draw is not only a simplified picture of a heavenly body, it is also a geometric shape as abstract as a circle or square. In Muslim architecture the star is purely abstract, seen in the complex geometric tracery of pierced screens or relief carvings. Its straight-sided, angular shape makes it an ideal motif for tiled floors. The complicated designs of encaustic tiles that survive in many English Victorian houses are another offshoot of Gothic Revival, and often include star shapes. In the same houses you may find a few panels of stained glass: a very popular design for the corner of a window was a little blue pane incised with a white, many-pointed star. These shapes recall the geometry of rock formations or snow crystals.

Above: Eight-pointed stars in the ceiling of the Mausoleum of Empress Galla Placidia, Italy, c440 BC.

Left: Twelfth-century Catalan fresco of a starry Archangel Michael weighing souls.

Above: The constellations in the dome of the old library of Salamanca University, Spain, painted by Fernando Gallego 1468-1507.

Stars have worked their way into many designs because of their use as a symbol. In heraldry, where they are called mullets, they signify the younger branch of a family, and often appear on crested silver, stained glass and porcelain. In America they are an outstanding motif of "colonial" style, associated above all with the beautiful quilts of the eighteenth and nineteenth centuries. Traditional designs include Blazing Star, Le Moyne Star, Virginia Star and Star of Bethlehem. It is not too hard to see a connection between these and the star-spangled banner which first flew in Baltimore in 1812, when it featured fifteen stars.

The patriotic star motif can also be found in early American stencils, used to decorate walls, floors and furniture. They were also used on tôleware, or painted tinware, which was popular

Above: Starry Sky over the Rhone, *Arles, September 1888, by Vincent van Gogh.*

throughout the nineteenth century and is enjoying a revival of interest now. A fine gold star motif is frequently used to decorate tôleware candlesticks, trays and cache-pots, and suits their crisply painted surface and elegant shapes very well. At Christmas, the pretty shape, gilded brightness and symbolic meaning of stars all combine to make them the perfect decoration.

It's worth thinking carefully about the number of points your stars will have when you are decorating with them, as this will affect the character of your design. Six-pointed stars are easy to draw as they can be marked out with a pair of compasses. Even numbers – six, eight, or twelve – are traditional in patchwork, and are often made out of triangles or diamonds in contrasting tones to give the motif a three-dimensional effect. A symmetrical star would be best for a project like a clock. You can make your design more elaborate by adding shorter points between the four or six main ones. A star with four points – a style often used to symbolize the star of Bethlehem, but also the basis for the logo of the 1951 Festival of Britain – usually has an elongated downward point: it is easy to exploit this idea to produce designs of great elegance.

Above: Symbolical depiction of the astrological breakthrough in the medieval conception of the world from Camille Flammarion, L'atmosphère météorologie populaire, *Paris 1888.*

The five-pointed star is the archetype. This is the star of the Stars and Stripes and the pentagram of the old alchemists. It can be drawn in five straight lines with one continuous stroke and its "endless knot" was credited with magical properties. It is the shape of the gold star that tells us we've achieved something wonderful. This is the one to use if you are scattering tiny stars over your ceiling or walls. Draw it freehand, deliberately irregular, for spontaneous, bold designs with a "handmade" feel. It has a chunky, squat shape, perfect for a naïve, unsophisticated look.

Real stars glitter. Make them shine with silver or gold. Use metallic paints, threads, shiny metal, foil and sequins and echo the night sky by setting them against soft deep blue or black. But experiment with paler colours too: delicate gold stars look stunning on a soft yellow or deep terracotta background. Leaving aside the real, glittery stars in the sky, you can follow the lead of traditional crafts and use this simple, well-defined shape as a focal point in folksy textiles and woodwork.

Have fun with these lovely projects and, when you've finished, give yourself a star.

Left: Stars in theatrical design – a set for The Magic Flute *at the Berlin Opera House in 1816.*

Above: Stars are inextricably linked with the Nativity as in this work by Master Francke, c1424.

Star Patchwork Herb Sachet

Patchwork stars made out of diamond shapes appear on many early American quilts; this one is based on the eight-point Lone Star motif. Lining the patches with paper is the traditional English way of making patchwork. It keeps the shapes sharp and accurate when joining the points of the star.

You Will Need

Materials
dark orange cotton fabric, 20 x 40 cm/8 x 16 in
mustard yellow cotton fabric, 12.5 x 20 cm/5 x 8 in
green and white check cotton fabric, 12.5 x 20 cm/5 x 8 in
matching sewing thread
dried herbs or pot pourri to fill
one small pearl button

Equipment
thin card for templates
thick paper
ruler
scissors
sewing needle
tacking thread
iron
pins

1 Using the templates, cut eight diamonds, four squares and four triangles from thin card. Adding a 5 mm/¼ in seam allowance: draw four yellow and four check diamonds, four orange squares and four orange triangles. Lay backing paper in the centre of each fabric shape, turn the seam allowances over the paper, folding neatly at the points, and tack in place.

2 Stitch a yellow and a check diamond together along one edge, then sew an orange square into the right angle. Make four of these units then join together to form a star. Sew the orange triangles into the remaining spaces to complete the square. Press lightly and remove all the tacking threads.

3 Cut a 19 cm/7½ in square from the remaining orange fabric and press under a 5 mm/¼ in seam allowance all round. With wrong sides together, pin this square to the patchwork and overstitch around the outside edge leaving a 7.5 cm/3 in gap on one side. Fill with herbs or pot pourri and sew up the opening. Sew the button to the centre of the star.

Appliqué Star Greetings Card

A beautiful birthday card to treasure in which the traditional craft of tin-punching is combined with appliqué and embroidery.

YOU WILL NEED

MATERIALS

check cotton fabric:
16 x 12 cm/6 x 4¾ in in light blue; 10 x 12 cm/ 4 x 4¾ in in medium blue; 4 x 16 cm/1½ x 6 in in dark blue
matching sewing thread
stranded embroidery thread in dark blue
metallic embroidery thread in silver
plain, unridged tin can
14 x 24 cm/5½ x 9½ in silver card, folded in half
all-purpose glue

EQUIPMENT

scissors
pins
sewing needle
can opener
tin snips
metal polish
thin card or paper for template
hammer
small piece of wood
bradawl or large nail

1 Cut the light blue fabric into four 4 cm/1½ in strips. Fold under the raw edge along the long side of each and pin to the medium blue fabric. Using blue embroidery thread, sew small running stitches close to the fold line on both edges.

2 Cut out four 4 cm/1½ in squares of dark blue fabric. Turn under the edges of each square and pin in each corner of the panel. Turn under the remaining edges and tack. Sew small running stitches in silver thread around the edges of the corner squares and embroider a simple star in the centre.

3 Remove the top and bottom of the tin can, cut down the back seam with tin snips and flatten out. Draw round the template three times. Cut the stars out carefully and hammer the points flat. Lay the stars right side up on the piece of wood and punch a star shape using the hammer and bradawl. Embroider three simple stars in silver thread. Glue the panel to the front of the card. Glue the tin stars in position.

Star Painted Candle Holder

The gentle glow of candles has an obvious affinity with starlight, and this twelve-pointed star is gilded and studded with copper to reflect the light. Painted in warm, festive colours, it would make a lovely addition to a traditional Christmas table.

You Will Need

Materials

5 mm/¼ in birch plywood sheet
1 cm/½ in pine
wood glue
white undercoat paint
acrylic paints in dark green, red, gold
matt varnish
six 2 cm/¾ in copper disc rivets

Equipment

pair of compasses
ruler
fret saw
sandpaper
paintbrushes
spike
wire cutters

1 Using the compasses, draw a large circle on the plywood. With the same radius, mark the six points of the star around the circle and join with a ruler. Draw a smaller circle on the pine and mark out the second star in the same way. Draw a circle in the centre to fit your chosen candle size.

2 Cut out the two star shapes and sand any rough edges. Stick together with wood glue to form a twelve-pointed star. Paint with white undercoat and sand lightly when dry. Cover with a base coat of dark green acrylic paint, then paint on the design. Seal with a coat of matt varnish.

3 Using a spike, make six holes for the copper disc rivets. Trim the stems of the rivets with wire cutters and push into the holes.

Dried Flower Star Wreath

In the spring the new, straight growth of woodland trees and shrubs is ideal for collecting and making into a wreath. Cut the twiggy stems early in the year before the buds have opened.

You Will Need

Materials
36 thin twigs approximately 60 cm/24 in long
matt acrylic paint in mid-blue
fine florist's wire
blue and white check ribbon 2 cm/¾ in wide, 3 m/3 yd
six pink dried or silk roses
selection of dried flowers and plant material in pink, purple and blue
all-purpose glue
small, variegated ivy leaves

Equipment
scissors
paintbrush

1 Trim the ends off the sticks to make them about 45 cm/18 in long. Paint each stick blue and allow to dry.

2 Overlap two bundles of six sticks and tie the ends securely with wire. Add a third bundle to complete the triangle. Weave the remaining bundles through the first triangle to make a six-pointed star and tie the ends with wire.

3 Wind the ribbon around the central hexagon and tie off neatly at the back. Build up the design, using the roses and small sprigs of pink, purple and blue plant material. Use glue to secure. Stick several ivy leaves at each corner of the hexagon.

4 Wrap a short piece of ribbon around five points of the star. On the sixth point, tie a longer piece into a loop.

STAR-SPANGLED BANNER

Make a bold statement with this cheerful wall hanging, based on an 1876 American bed quilt.

YOU WILL NEED

MATERIALS
fusible bonding web, 30 x 40 cm/12 x 16 in
blue cotton fabric, 76 x 142 cm/30 x 56 in
white cotton fabric, 76 x 40 cm/30 x 16 in
dark red cotton fabric, 76 x 38 cm/30 x 15 in
medium weight wadding, 66 x 66 cm/26 x 26 in
curtain pole with decorative finials, 86 cm/34 in
red cord, 130 cm/50 in
six 2.5 cm/1 in curtain rings
acrylic paint in dark red and white
matching sewing thread

EQUIPMENT
tracing paper, pencil
thin card or paper for template
scissors
iron
pins
sewing machine
tacking thread or safety pins
sewing needle
paintbrush

1 Trace and enlarge the star template at the back of the book, and transfer it onto the fusible bonding nine times. Cut out the stars, iron onto white fabric, then cut out. Peel off the backing paper. Cut nine 14 cm/5½ in blue cotton squares and fuse a star to the centre of each, making sure all the stars lie in the same direction.

2 Neaten the edges of the stars by stitching over them with a narrow satin stitch in white thread.

3 Cut out four red and four white rectangles, 7.5 x 14 cm/3 x 5½ in. Press all seams flat and join the red and white pieces in pairs along the longer sides. Lay the resulting squares alternating with the blue star squares in a chequerboard pattern to form a border around the central star. Pin and sew together in three rows of three, then join the rows to form a square. Cut out four red and four white rectangles, 7.5 x 40 cm/3 x 15½ in, and join in pairs along the longer sides. With right sides together stitch a blue square to each end of two of the rectangles. Stitch the remaining pieces to opposite sides of the central panel, with the white sides on the inner edge, then stitch the longer strips to the other two sides.

4 Cut 66 cm/26 in squares of wadding and blue cotton and baste to the patchwork with tacking or safety pins. Machine or hand quilt along the seam lines, then stitch all round the outside, 3 mm/⅛ in from the edge. Trim.

5 From the remaining blue cotton cut four 3 x 66 cm/1¼ x 26 in strips for the binding. Press in half lengthways, then press under 5 mm/¼ in along one edge. Pin each strip along one side of the quilted square, raw edges even, and stitch. Turn the folded edge to the back and slipstitch in place. Neaten the corners.

6 Sew the curtain rings to the top of the banner, spacing them evenly.

7 Remove the finials from the curtain pole and paint with dark red paint, using a dry brush for a dragged effect. Use cream paint to pick out stripes or fine details on the turned ends. Thread the pole through the curtain rings and replace the finials. Attach the cord to one end of the pole and wrap it round several times, securing with matching cotton. Do the same with the other end, then make a loop in the centre.

Shooting Star Badge

This jolly little shooting star will brighten up a plain jumper or jacket. Use pearlized paint for its tail and glossy varnish to make the colours glow.

YOU WILL NEED

MATERIALS
small piece of 4 mm birch plywood
white undercoat paint
acrylic paints
water-based pearlized paints
gloss varnish
all-purpose glue
brooch pin

EQUIPMENT
tracing paper
pencil
fret saw
sandpaper
paintbrushes

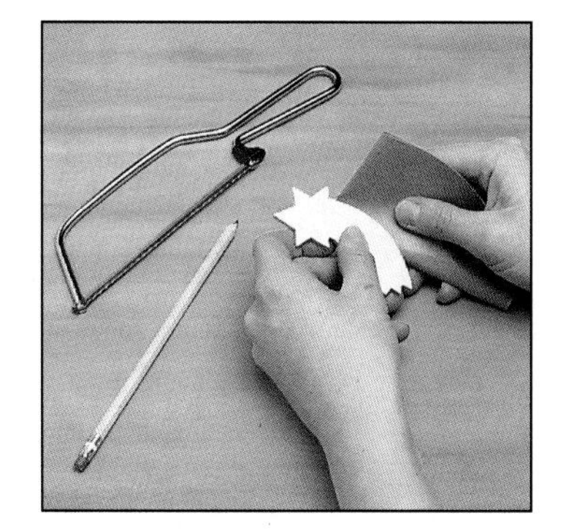

1 Trace the template at the back of the book and transfer to the plywood. Cut out. Sand all the edges.

2 Paint with a coat of white undercoat. When dry, sand lightly and mark the remaining points of the star in pencil.

3 Paint on the badge's design in acrylic paints, using pearlized paint for the tail. Protect with a coat of gloss varnish.

4 Glue the brooch pin to the back of the badge.

A Star for the Christmas Tree

Persuade the fairy to take a well-earned rest this year, and make a magnificent gold star to take pride of place at the top of the Christmas tree.

You Will Need

Materials
corrugated cardboard
newspaper
PVA glue
gold spray paint
gold relief/puff paint
gold glitter
thin gold braid

Equipment
tracing paper
pencil
thin card or paper for template
scissors
metal ruler
craft knife
bowl
paintbrush

1 Trace the template from the back of the book onto thin card or paper. Cut it out and draw round it on the corrugated cardboard. Cut out using a craft knife and metal ruler. Tear the newspaper into small strips. Thin the PVA glue with a little water and brush it onto both sides of the newspaper, coating it thoroughly. Stick it on the star, brushing it down with more PVA to get rid of air bubbles. Work all over the star in a single layer, covering the edges and points neatly. Allow to dry, then apply a second layer.

2 If the star begins to buckle, place under a heavy weight. When completely dry, spray both sides gold and allow to dry.

3 Draw the design on one side in gold relief paint and sprinkle with glitter while still wet. Allow to dry completely before repeating the design on the other side. Attach thin gold braid with which to hang the star from the top of the tree.

Star Batik T-shirt

The lovely Indonesian craft of hand-painted batik is used here to transform a plain T-shirt with a simple but striking star motif. Using a traditional wax pen called a tjanting, the design is drawn in wax. Take care when heating the wax as it ignites if it gets too hot.

YOU WILL NEED

MATERIALS
white T-shirt
batik wax
batik dyes in brilliant red, peacock blue and royal blue

EQUIPMENT
thin card or paper
black felt pen
quilting pencil or charcoal
batik or quilting frame
newspaper
batik wax pot
tjanting
paintbrush
wooden spoon
washing soda
iron

1 Enlarge the design at the back of the book to suit the size of the T-shirt. Position the pattern inside the T-shirt and trace onto the fabric. Fit the frame on the front of the T-shirt and put layers of newspaper inside to protect the back.

2 Heat the wax until hot enough to seep into the fabric. With the tjanting, draw a line of wax around the design. Fill in the details that are to stay white. Paint some areas with the red dye solution and then allow to dry flat.

3 Wax over the parts that are to stay red. Repeat the process with the blue dyes. Once the design is complete, allow to dry flat. Paint wax over both sides of the design to seal. Make a royal blue dye bath in a large bucket. Immerse the T-shirt in the bath and stir continuously for six minutes. Add 15 ml/1 tbsp of washing soda dissolved in warm water. Leave the shirt to soak for 45 minutes. Rinse and dry flat. Iron over sheets of newspaper to melt excess wax.

Astral Clock

An impressive starburst clock with rays cut out of shining copper. The face is a simple disc of clay, painted in turquoise to give a verdigris effect.

YOU WILL NEED

MATERIALS
0.5 mm copper sheet, 10 x 25 cm/4 x 10 in
self-hardening modelling clay, 450 g/1 lb
acrylic paints in deep turquoise, lemon yellow and white
varnish
gold powder
epoxy resin glue
clock workings, hands and battery

EQUIPMENT
tracing paper
pencil
thin card or paper for templates
metal cutters
sandpaper
rolling pin
modelling tools
paintbrush

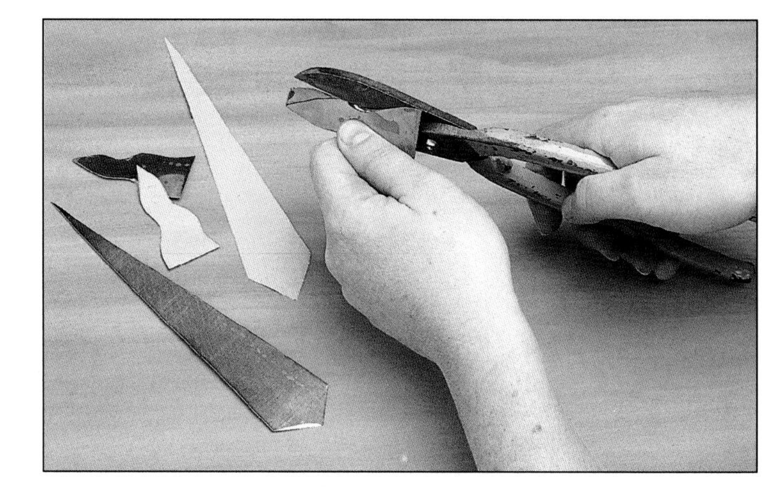

1 Trace the templates at the back of the book and cut out of card or paper. Draw the outlines for four large and four small rays on the copper sheet. Cut out with metal cutters and sand the edges.

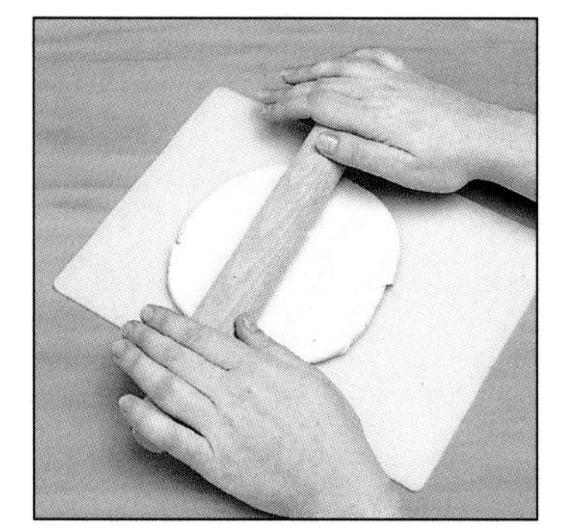

2 Roll out the clay to a flat sheet 5 mm/¼ in thick.

3 Place the template on the clay and cut out the shape. Trace the inner circle with a modelling tool to impress the shape in the clay.

4 Roll a clay snake and place along the traced inner circle, joining the ends. Use a modelling tool to join it to the clock face and smooth with wet fingers. Make a hole in the middle of the face.

5 Press the short rays into the side of the clay and leave to dry completely.

6 Mix turquoise, yellow and white paint and paint the face of the clock.

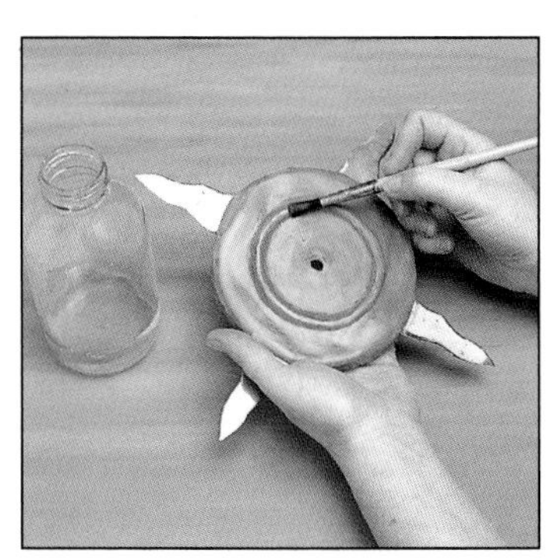

7 Paint the face with a layer of varnish and leave to dry.

8 Mix the gold powder with varnish and decorate the ridge around the clock with thick gold lines.

9 Bend the bases of the long copper rays to fit over the edge of the face. Fix firmly in position with epoxy resin glue.

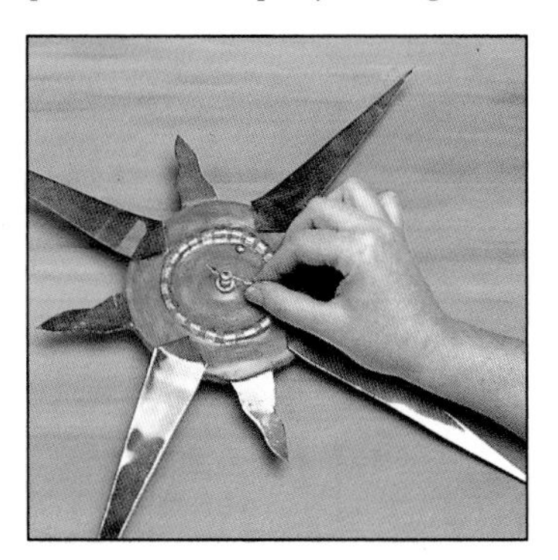

10 Fix the mechanism to the back of the clock and screw the hands onto the front. Insert a battery.

Starry Letter Rack

Painted in a lovely midnight blue with jolly yellow stars, this charming letter rack is quite straightforward to assemble.

You Will Need

Materials

5 mm/¼ in birch plywood sheet
wood glue
four x 15 mm/⅝ in wooden balls
white undercoat paint
acrylic paints
satin varnish

Equipment

tracing paper
pencil
ruler
pair of compasses
fret saw
sandpaper
masking tape
paintbrushes

1 Trace and enlarge all the pieces from the back of the book. Mark on the plywood. Cut out and sand all edges.

2 Glue the pieces together and hold in place with masking tape until the glue has hardened completely.

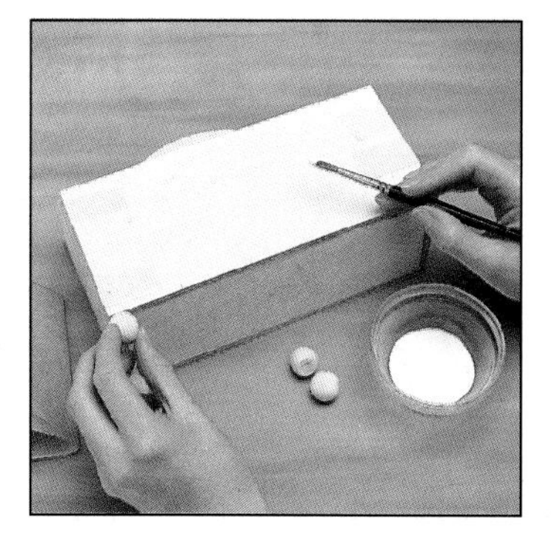

3 Remove the tape and sand all the edges and corners. Glue the wooden balls to the corners of the base.

4 Paint on a coat of white undercoat, sanding down lightly when dry. Paint the rack and the stars in acrylic paints. Seal with a coat of satin varnish. When the varnish is completely dry, glue the stars in position on the front of the rack.

Wooden Star Picture Frame

Make your favourite person an instant star by putting their picture in this original frame. Once the photograph is in position, back it with a piece of stiff card cut to size and held in place with masking tape.

You Will Need

Materials
5 mm/¼ in birch plywood sheet
white undercoat paint
acrylic paints
satin varnish
brass triangle picture hook and pins

Equipment
ruler
pair of compasses
pencil
fret saw
sandpaper
router
paintbrushes
hammer

1 Measure your picture and subtract 1 cm/½ in from each dimension to work out the size of the frame opening. Mark this on the plywood, then mark out the star pattern from the back of the book around it.

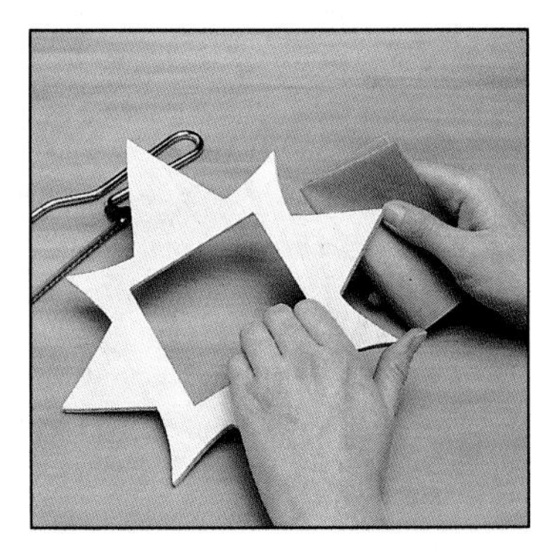

2 Cut out the star shape and inner square and sand down all the edges. On the back, rule a line round the opening 5 mm/¼ in from the edge and make a rebate with the router 3 mm/⅛ in deep.

3 Sand down then paint the frame with white undercoat, sanding down lightly when dry. Mark out the bands in pencil then paint the design.

4 Finish with a coat of satin varnish. Finally, pin a triangle hook to the top point of the frame.

Jewelled Star Tree Ornaments

You can buy pre-cut polystyrene shapes from craft suppliers, and these little stars are ideal for dressing up to go on the Christmas tree. Multi-coloured sequins and tiny beads make them into exotic treasures. Attach the sequins along the points of the stars or cover the whole surface for an extravagant effect.

YOU WILL NEED

MATERIALS
polystyrene star shape, approximately 8 cm/3 in across
gold spray paint
multi-coloured glass seed beads
a few seed pearl beads
multi-coloured sequins
a few special design sequins
1.5 cm/½ in brass-headed pins
thin gold braid

EQUIPMENT
small piece of plasticine

1 Spray the polystyrene shape gold, anchoring it with a piece of plasticine to stop it blowing away when spraying. Allow to dry.

2 Sort the beads, sequins and pins into different containers to make it easier to choose your colours and shapes as you work.

3 Thread a seed bead onto a pin, followed by a sequin. Push gently into the polystyrene. Repeat to complete the design. In the centre of each side of the star pin a seed pearl and a cup sequin or other special design.

4 Finally, attach a small loop of thin gold braid to hang from the tree.

STAR-SPANGLED VELVET SCARF

A lavish scattering of gold appliqué and sparkling beads on sumptuous dark velvet creates a luxurious scarf for winter evenings. Arrange the beads in clusters around the large stars to appear like stardust.

YOU WILL NEED

MATERIALS
burgundy velvet, 23 x 63 cm/ 9 x 25 in
gold velvet, 23 x 63 cm/9 x 25 in
fusible bonding web, 23 x 30 cm/ 9 x 12 in
gold machine embroidery thread
translucent gold rocaille beads
matching sewing thread
black velvet, 32 x 122 cm/ 12 ½ x 48 in
black glazed cotton for lining, 56 x 89 cm/22 x 34½ in

EQUIPMENT
tracing paper
pencil
scissors
thin card or paper for templates
iron
pressing cloth
sewing machine
fine sewing needle
pins

1 Cut the burgundy velvet into two rectangles each measuring 23 x 32 cm/ 9 x 12 ½ in. From the gold velvet cut two 4 x 32 cm/ 1½ x 12½ in strips and two 6 x 32 cm/2½ x 12½ in strips. Trace and enlarge the star templates and draw each star twice on the fusible bonding. Cut out roughly and iron onto the wrong side of the remaining gold velvet, then cut out each star neatly along the outline. Peel off the backing paper and arrange eight stars on each burgundy rectangle. Iron in place using a pressing cloth. Using gold thread machine around the edge of each appliqué star and work a spiral over the centres of the three largest shapes. Sew a thick sprinkling of beads to the background with double thread.

2 Join one wide and one narrow gold velvet strip to the long sides of each burgundy panel, using a 1 cm/½ in seam allowance. Attach a panel to each end of the black velvet, joining the narrow gold strip to the main scarf. Press all seams open lightly, using a pressing pad and pressing cloth. Cut the lining fabric in half lengthways and join to form one long strip. Press the seam open, then pin the lining to the scarf along the long edges with right sides facing. Stitch, leaving a 12.5 cm/ 5 in opening in the centre of one seam. Remove the pins then adjust the ends so that an equal amount of velvet lies on each side of the lining. Pin, then stitch right across the ends. Clip the corners and turn the scarf to the right side. Press lightly and slip-stitch the opening.

Spice Scented Star Pot Stand

The lovely homespun look of this pot stand is achieved by tinting all the fabrics with tea. Placing a hot pot on the mat releases a rich, spicy scent of cloves.

YOU WILL NEED

MATERIALS
calico, 18 cm/7 in square
red gingham, 22 x 44 cm/9 x 18 in
blue ticking, 6 x 22 cm/2½ x 9 in
small blue check cotton, 6 x 22 cm/2½ x 9 in
tea bags
stencil crayon in yellow ochre
stranded embroidery threads in yellow ochre and beige
four odd buttons
sewing thread
whole cloves

EQUIPMENT
bowl
iron
stencil card
craft knife
ruler
masking tape
stencil brush
kitchen paper
pins
sewing needle
sewing machine

1 Wash all the fabrics to remove any dressing. Brew some strong tea and soak the fabrics until you are satisfied with the colour. It is best to do this in stages, re-dipping if you need to make them darker. Leave to dry and press well.

2 Trace the star template at the back of the book and transfer it to the stencil card. Cut out the star using a craft knife and a ruler.

3 Place the stencil in the centre of the calico square and fix with masking tape. Work around the card with the stencil crayon, scribbling the stencil paint near the edges of the shape, avoiding getting any on the fabric. Now work the stencil brush into the crayon and gently ease the paint from the stencil onto the fabric with a light scrubbing action. Add more paint if necessary. Do not try to get a completely even coverage as this adds to the "antique" effect.

4 Iron the calico on the wrong side between sheets of kitchen paper to fix the motif and blot excess paint. Fold under the edges of the calico until it measures 11.5 x 12 cm/ 4½ x 4¾ in. Press.

5 Pin the calico in the centre of one gingham square. With your fingers, gently fray one long edge of the ticking and check strips and pin them to opposite sides of the gingham, with the frayed edges pointing inwards.

6 Using three strands of yellow embroidery thread, stitch the strips to the gingham with a simple running stitch near the frayed edges. Using beige embroidery thread and running stitch, attach the calico square; sew a button in each corner. With wrong sides together, machine the second gingham square to the decorated square, leaving an opening in one side. Turn to the right side.

7 Fill with whole cloves. Do not overfill or the pot will be unsteady when resting on the mat.

8 Neatly sew up the opening by hand.

Celestial Mobile

Stars, the crescent moon, and a huge shooting star jostle each other in this amusing mobile. The little star-shaped cut-outs in the meteor's tail can all be joined by saw cuts, so you need drill only one hole through which to pass the saw blade.

You Will Need

Materials
4 mm birch plywood
white undercoat paint
chrome finish spray paint
acrylic paints
varnish
0.14 mm nylon fishing line

Equipment
fret saw
drill
sandpaper
paintbrushes

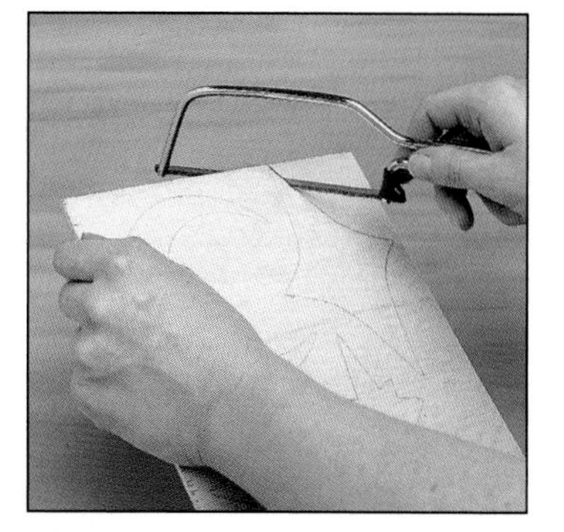

1 Trace the templates from the back of the book and transfer them to the plywood. Cut them out using a fret saw. Drill a small hole inside the circle of the shooting star and through the first cut-out in the tail. Pass the saw blade through each of these to make the internal cuts. With a very small bit, drill holes through the top of each piece and in the marked positions on the hanger. Lightly sand all the pieces.

2 Paint all the pieces with undercoat. Allow to dry, then spray the top of the shooting star and the crescent moon with chrome paint. Complete the decoration using acrylic paints and varnish.

3 Tie the stars to the hanger with thin nylon line. Make sure they do not bump into one another. Add a loop of nylon to the hole at the top.

Gothic Star Mirror

Use curly sausages of clay and cut out simple, freehand star shapes to decorate the frame of your mirror. To make sure the frame is symmetrical, fold a sheet of paper in half, draw a simple template and transfer it to the cardboard before you cut.

You Will Need

Materials
thick cardboard
mirror, 12.5 x 7.5 cm/5 x 3 in
masking tape
short length of thin wire
newspaper
PVA glue
white emulsion
self-hardening clay
acrylic paints
varnish

Equipment
thin card or paper for template
craft knife
paintbrushes
modelling tools

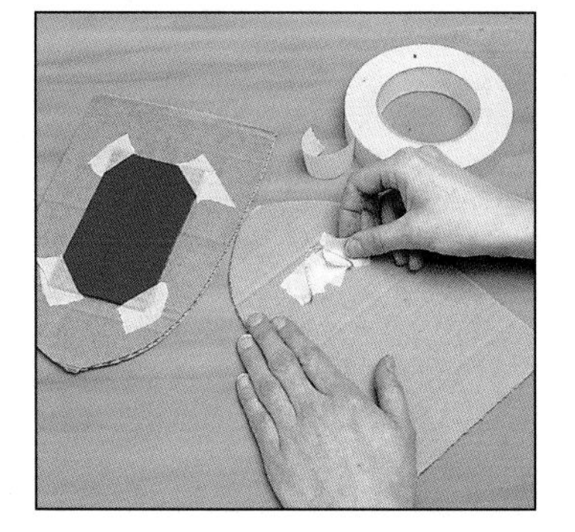

1 Draw the shape of the frame onto the cardboard twice and cut out with a craft knife. Cut out the central shape from the front piece of the frame. Fix the mirror to this piece with masking tape. Make a hook from wire and fix it securely to the back section of the frame. Put the two pieces of card together, sandwiching the mirror, and tape securely.

2 Tear the newspaper into squares approximately 2.5 cm/1 in across and use PVA glue to stick them in a single layer on both sides and around the edges of the frame. Leave until completely dry.

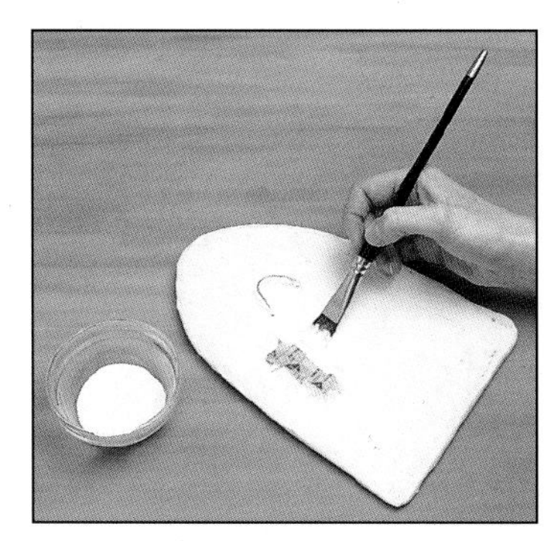

3 Prime with white emulsion and allow to dry. Roll some clay into long sausages and use to make an edging for the mirror and decorative scrolls and curls. Press onto the frame with the help of a modelling tool. Cut star shapes out of the clay and fix on. Leave to dry. Decorate the frame with acrylic paints and varnish when dry.

Muslin Star Curtain

Felt stars are caught in the deep hem of this unusual, sheer curtain. When machining them in, match the sewing thread to each of the colours of the felt. Add 36 cm/14 in to the depth of your window for the hem and heading when calculating the length of the muslin.

You Will Need

Materials

felt squares in purple; light, medium and dark blue; light and dark pink
white muslin to fit window (see above)
matching sewing threads
white sewing thread
curtain tape

Equipment

tracing paper
pencil
thin card or paper for template
scissors
tailor's chalk
iron
pins
tacking thread
sewing needle
sewing machine

1 Trace the star motif at the back of the book, transfer to thin card and cut out. Drawing round the template with tailor's chalk, draw two stars on each colour felt. Cut out.

2 Make a 25 cm/10 in hem in the muslin and press along the fold. Open it out and arrange the felt stars above the fold line and within the depth of the hem.

3 Carefully pin the hem back over the stars, turning under 1 cm/½ in along the raw edge. Pin the stars in position and tack round the edges 5 mm/¼ in outside the felt.

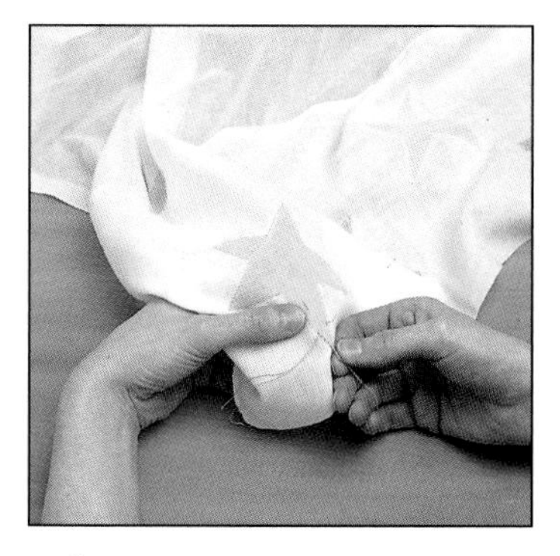

4 Machine stitch around each star using matching sewing thread. Sew the ends under the muslin and trim. Stitch the hem and add curtain tape at the top of the curtain.

Moulded Star Earrings

A clay mould is used to model these eye-catching silver earrings, so it's easy to make as many pairs as you want – as gifts for everyone who admires them on you!

You Will Need

Materials
self-hardening modelling clay
pair of earring studs
all-purpose glue
black acrylic paint
silver powder
varnish

Equipment
rolling pin
tracing paper
pencil
thin card or paper for template
modelling tools
paintbrushes

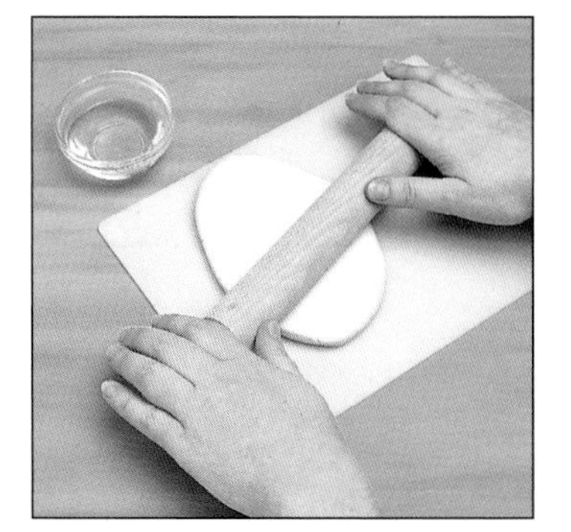

1 Roll out a small piece of clay to a thickness of 8 mm/⅜ in.

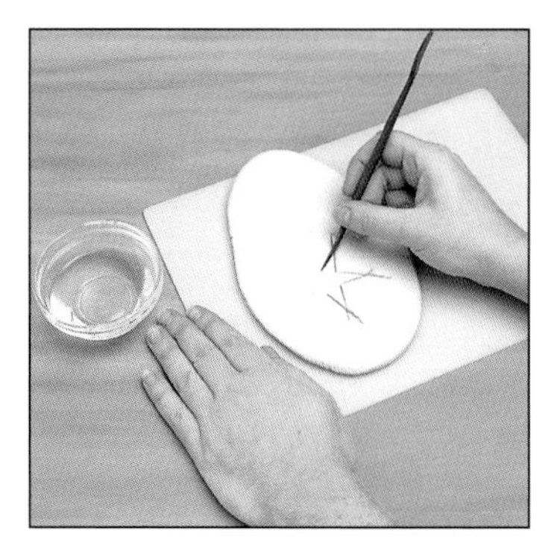

2 Trace the template from the back of the book onto thin card or paper. Cut the star shape out of the clay.

3 Mark a line from the centre of the star to each point where two rays meet and use the flat side of the modelling tool to mould each point to a 90° angle. Smooth the star with water, tuck the edges in neatly and leave to dry.

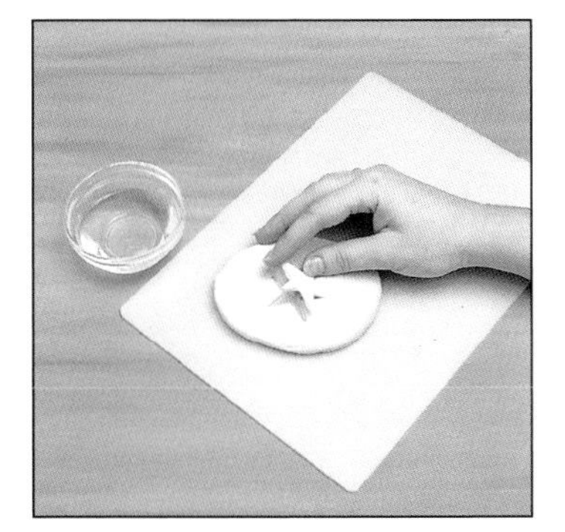

4 Take a small ball of clay and press with your palm until it is about 2 cm/¾ in thick. Press in the hardened clay star then lift out carefully without distorting the mould. Leave to dry.

5 Use the mould to make further clay stars. Lift them out of the mould and place face up on the work surface.

6 Trim off the excess clay with a modelling tool. Allow to harden.

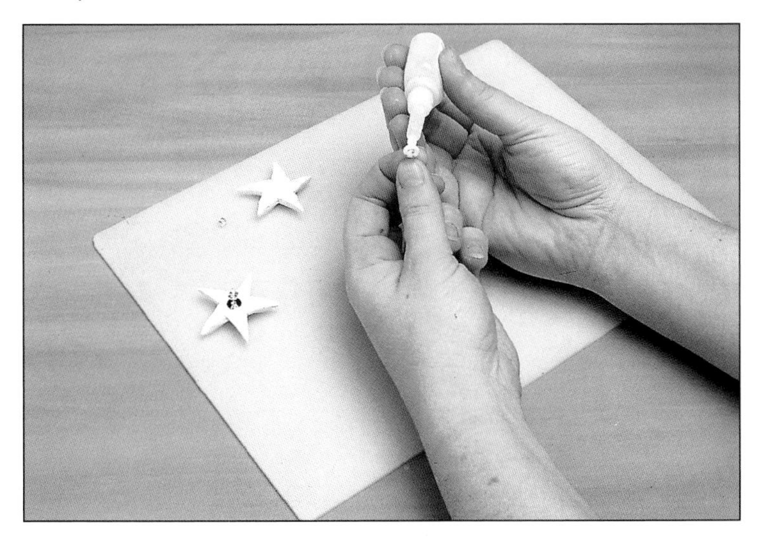

7 Glue earring studs to the backs of the stars.

8 Paint the stars with black acrylic paint and leave to dry completely.

9 Mix silver powder with varnish and brush this over the stars to complete.

Gilded Star Stained Glass Bottle

Just a corner of a star forms the motif on this lovely, glowing bottle. Use a square-shaped bottle or jar so that you can leave it lying flat while the paint dries completely before you begin work on the next side.

You Will Need

Materials
flat-sided glass bottle or jar
methylated spirit
gold glass-painting outliner
solvent-based glass paints in red, blue, green, yellow

Equipment
kitchen paper
paper
paintbrushes

1 Wash the bottle thoroughly in hot water and detergent, then wipe with kitchen paper and methylated spirit to remove all traces of grease.

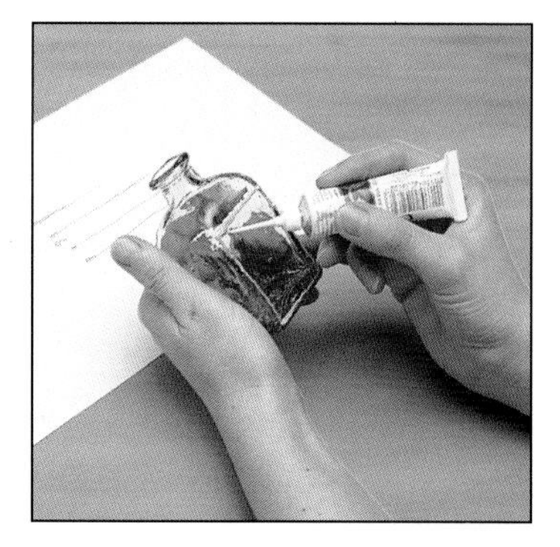

2 Lay the bottle on its side. Practise with the gold outliner on a piece of paper first, then draw on the design from the back of the book. Allow to dry for about 12 hours.

3 Apply the glass paint between the outlines, brushing it on thickly to avoid streaky brush strokes. Leave the bottle lying on its side and allow to dry for at least 36 hours before starting to paint the next side.

Star Print Wrapping Paper

Complete an original gift by dressing it up in original, hand-decorated wrapping paper. It's not only pretty, but fun to print. This star pattern printed in festive colours makes great Christmas wrapping.

YOU WILL NEED

MATERIALS
thin card
plain wrapping paper
white chalk
water-based block printing paint in red, green and white
water-based gold paint

EQUIPMENT
scissors
rubber stamp with star motif
paintbrush

1 Cut a circular template out of card and draw around it in chalk on the wrapping paper, spacing the circles evenly on the sheet.

2 Print alternate circles with red and green stars, brushing the paint evenly on the stamp between each print.

3 Print white stars in the middle of each circle, between the circles and in each corner.

4 Following the chalk circle between the stars, make rings of gold dots and dot the point of each star with gold.

TEMPLATES

To enlarge the templates to the correct size, use either a grid system or a photocopier. For the grid system, trace the template and draw a grid of evenly spaced squares over your tracing. To scale up, draw a larger grid onto another piece of paper. Copy the outline on to the second grid by taking each square individually and drawing the relevant part of the outline in the larger square. Finally, draw over the lines to make sure they are continuous.

Star Patchwork Herb Sachet p12

Star Batik T-shirt p28

Star-spangled Velvet Scarf p40

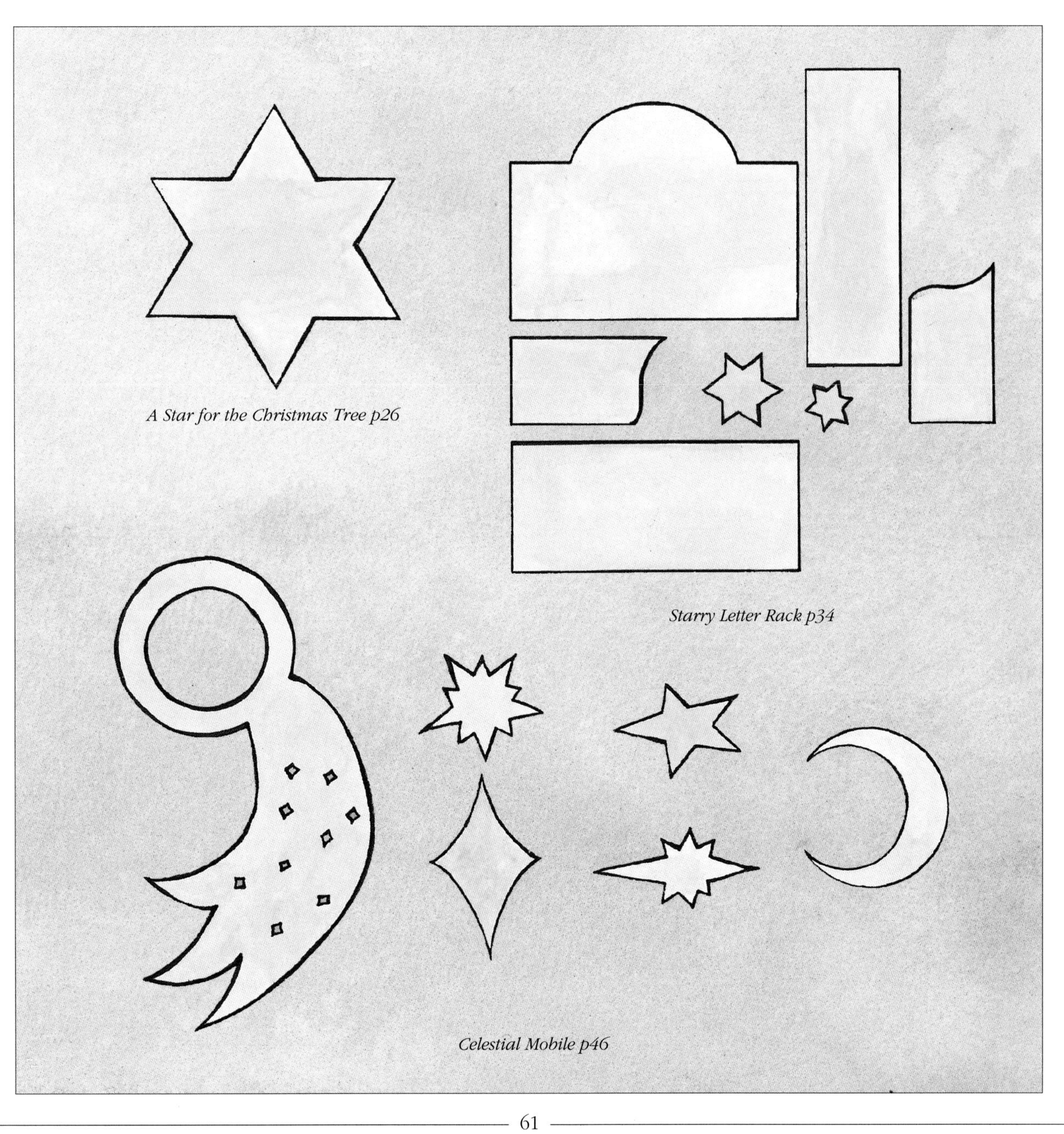

A Star for the Christmas Tree p26

Starry Letter Rack p34

Celestial Mobile p46

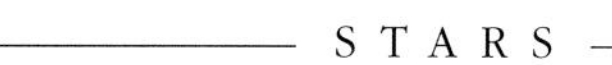

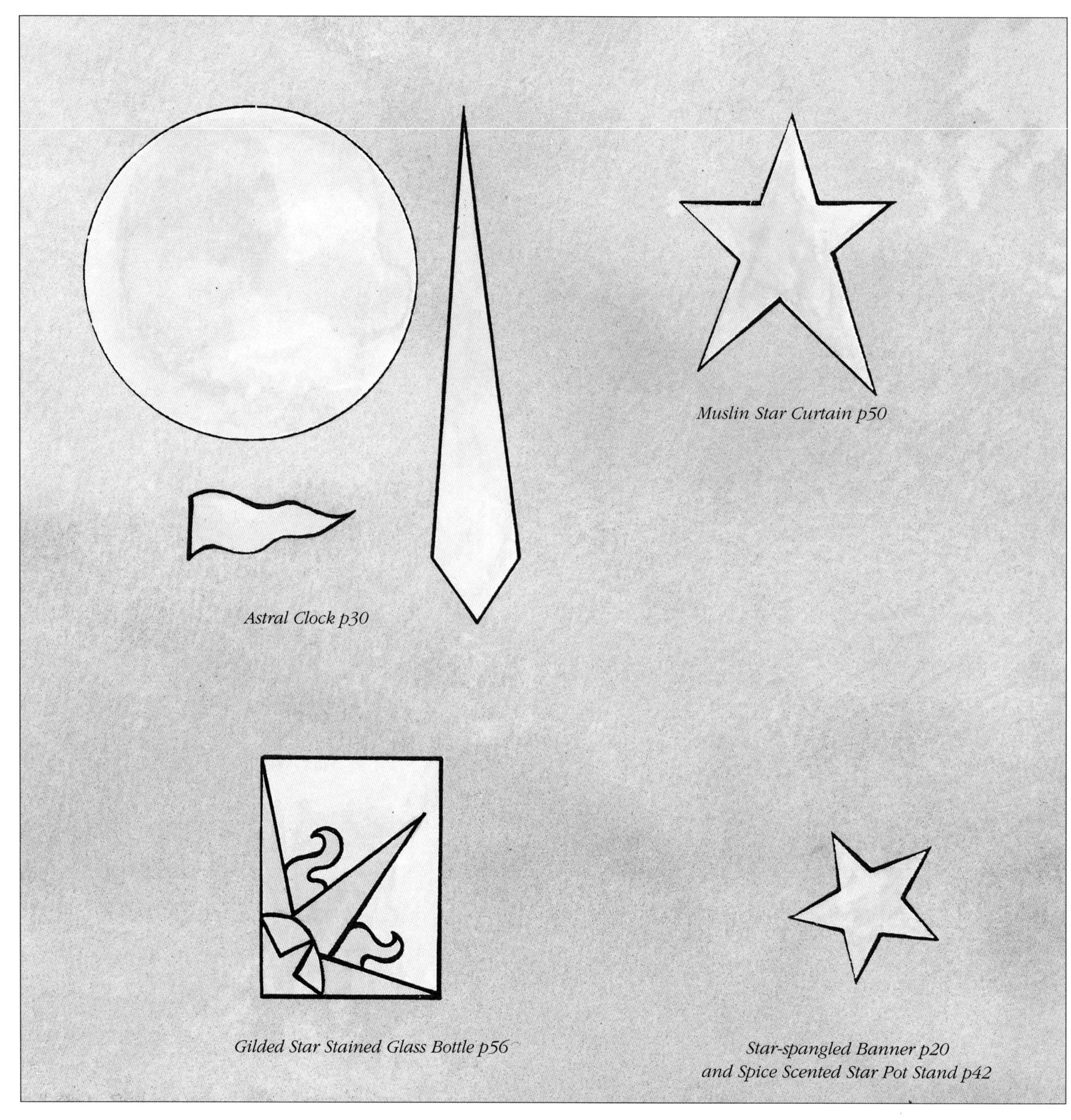
Muslin Star Curtain p50
Astral Clock p30
Gilded Star Stained Glass Bottle p56
Star-spangled Banner p20
and Spice Scented Star Pot Stand p42

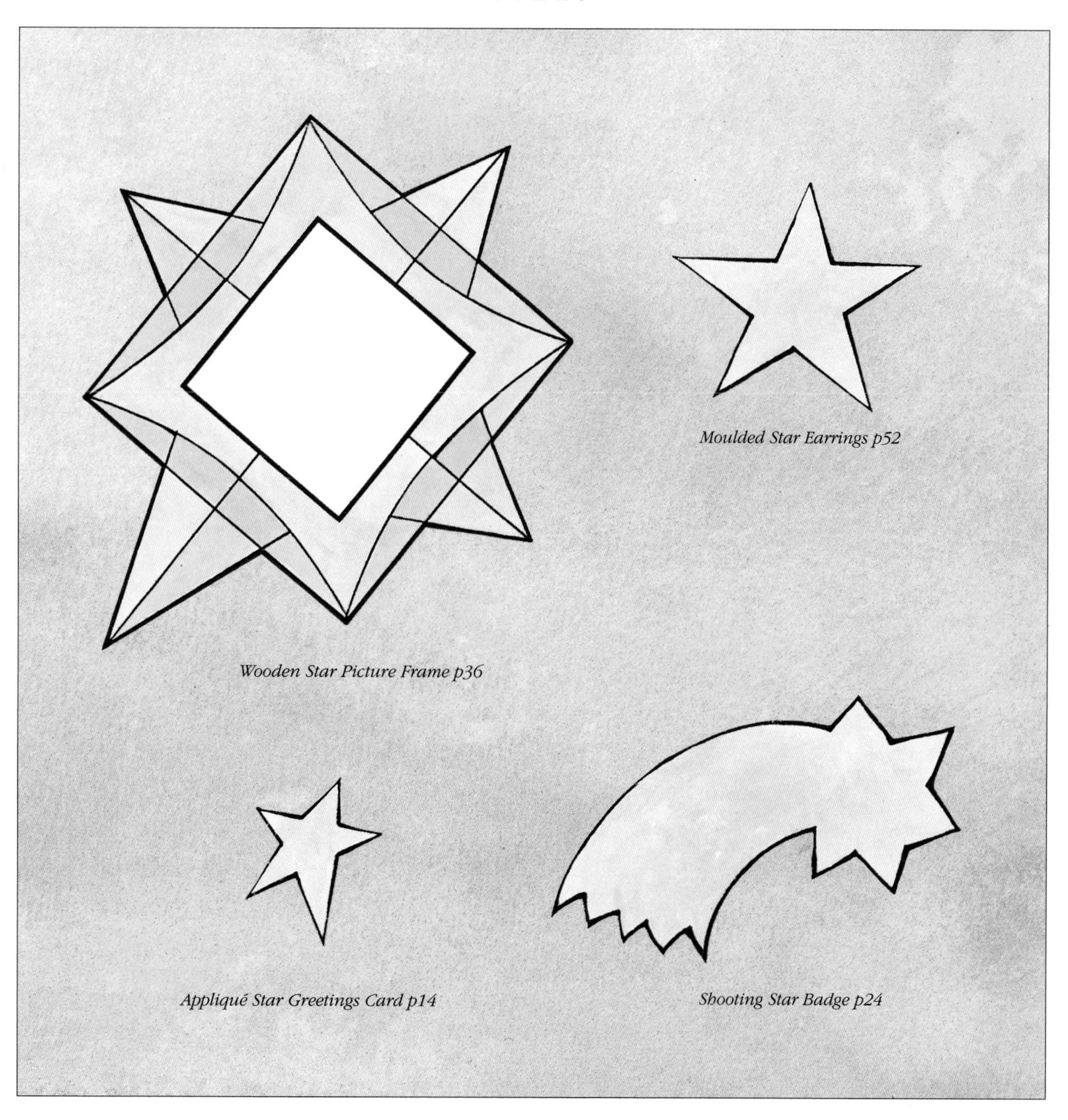

Moulded Star Earrings p52

Wooden Star Picture Frame p36

Appliqué Star Greetings Card p14

Shooting Star Badge p24

ACKNOWLEDGEMENTS

The author and publishers would like to thank the following people for designing the projects in this book:

Ofer Acoo

Astral Clock p30;
Moulded Star Earrings p52

Amanda Blunden

Gothic Star Mirror p48

Penny Boylan

Jewelled Star Tree Ornaments p38;
A Star for the Christmas Tree p26;
Spice Scented Star Pot Stand p42

Lucinda Ganderton

Star-spangled Banner p20;
Star-spangled Velvet Scarf p40;
Star Patchwork Herb Sachet p12

David Hancock

Celestial Mobile p46

Jill Hancock

Starry Letter Rack p34;
Wooden Star Picture Frame p36;
Star Painted Candle Holder p16;
Shooting Star Badge p24

Emma Petitt

Gilded Star Stained Glass Bottle p56

Emma Whitfield

Star Print Wrapping Paper p58

Dorothy Wood

Appliqué Star Greetings Card p14;
Dried Flower Star Wreath p18;
Star Batik T-shirt p28;
Muslin Star Curtain p50

Picture Credits
The Publishers would like to thank Photo AKG London for supplying the pictures on pages: 8, 10 and 11, and to E.T. Archive for the picture on page 9.